LOOKING AT SCIENCE 4

The Physical World

David Fielding

Basil Blackwell

© 1984 David Fielding

All rights reserved. No part of this publication may be reproduced, stored in a retrieval system, or transmitted in any form or by any means, electronic, mechanical, photocopying, recording or otherwise, without prior permission of Basil Blackwell Publisher Limited.

First published 1984

Published by Basil Blackwell Limited
108 Cowley Road
Oxford OX4 1JF

ISBN 0 631 91380 7 (*paperback*)
    0 631 13697 5 (*cased*)

Printed in Hong Kong

Topic symbols

■ This work is about air and water.

◆ This work is about animal life.

◆ This work is about electricity and magnetism.

◐ This work is about light and dark.

▲ This work is about mechanics.

✿ This work is about plant life.

🌱 This work is about weather and climate

Look for the symbols in the other books in the series. There is more work about these things in the other books.

Contents

A word to teachers and parents   5

Part 1   Weather and survival

Wind   6
What an anemometer is. How anemometers measure the force of the wind. How winds are made.

A weather vane   8
How we can find the exact direction of the wind. How weather vanes work, and why they are placed high above the ground.

Keeping warm   10
How warm things lose their heat. How we keep warmth in. How animals and humans are insulated.

Conquering the cold   12
Using electricity to make heat and light.

Rain   14
Measuring the amount of rain that falls. Why we do this. How rain is made.

Safe in bad weather   16
What a magnetic compass is, and how it helps us find our way. How it works.

Weather and survival   18
How people have learned to survive bad weather. Weather forecasts, and how barometers help us to make them.

Part 2   Light and how we use it.

How we see   20
How eyes and cameras work.

Reflections   22
What mirrors are, and how they make reflections.

Seeing round corners   24
How mirrors can let us see round corners. How a periscope works.

Strange reflections   26
Why curved mirrors produce strange reflections. Why specially curved mirrors can be useful.

Magnifying   28
Glass bends rays of light. How we can use glass to make things look bigger.

Coloured light   30
Light can be of different colours. These colours can be mixed. How colour television works.

Light and how we use it   32
About photography, lenses and telescopes

Part 3 Magnetism and electricity

 Magnetism from electricity   34
Electricity can produce magnetism. What electromagnets are, and how they work.

 Sound from electricity   36
How to use a switch and an electromagnet to make a series of sounds.

 Electricity and safety   38
Things which let electricity flow through them. Things which block electricity. Insulation.

 Switches   40
How switches work, and how we use them to control the flow of electricity. Switches that let us send messages in code.

 Electricity from a magnet   42
Making an electric current by moving a magnet in a coil of wire. How we use this idea to make generators.

 Weak and strong electricity   44
The brightness of bulbs can be changed according to how they are arranged in a circuit. Bulbs and batteries in series and in parallel.

 Magnetism and electricity   46
How telegraphs, microphones, earphones and loudspeakers work.

New words   48

Acknowledgements

Avitec 31(3)
Barr and Stroud Limited 24(1)
Central Electricity Generating Board 38(1)
Geoslides Photo Library 11(2)
Sally and Richard Greenhill 26(1)
Griffin and George 43(3)
North of Scotland Hydro-electric Board 43(2)
Brian Peart 32(1)
Ken Pilsbury 9(3)
Audrey Scott/Channel Swimming Association 11(3)
Rebecca Skillman 28(1)
Topham Picture Library 33(3), 47(2)
ZEFA 23(2)

Illustrations by Michael Stringer (colour)
and David Fielding (black and white)
Design by Indent, Reading

A word to teachers and parents

The Physical World is the fourth of five books in the *Looking at Science* series. The series has been designed to do two things:

- It gives children a solid body of knowledge in natural and physical science.

- It begins to introduce them to the nature of scientific enquiry.

These two elements are developed side by side through the books.

Each double page covers a particular area for study. The left hand page outlines an activity to perform and the right hand page gives information connected with it. Each book also contains another kind of double page, which is purely factual, spaced at regular intervals. These pages draw together the themes of the preceding pages.

The activities are introduced with the symbol ♥, and cover experimentation, observation and recording. A list of all the equipment needed for the experiments is given near the beginning of each spread.

These books can be worked through in order. Alternatively, they can be used as source material for topic work. Suggested topic areas are identified, with symbols, in the contents list.

The Physical World looks at three areas where science has a definite effect on our lives.

Part 1   Weather and survival
This section looks at what causes weather, how it is forecast and recorded, and at how people cope with bad weather. It investigates the devices used for measuring wind and rain, and why we need the information. It goes on to look at insulation – natural and man-made, and of how people use other forces, such as electricity and magnetism, to overcome the problem caused by weather.

Part 2   Light and how we use it
We see the world around us through our eyes which receive and use the light reflected off objects. This section discusses the structure and working of the eye and goes on to investigate how light is reflected by different surfaces. People have discovered and developed ways of bending and intensifying light rays, for example, magnifying lenses, and the section looks at how these facts are applied in practice.

Part 3   Magnetism and electricity
Magnets can be used to create electricity and vice-versa. This section looks at why it can be useful to use one force to create the other, and at how the electromagnets and electricity are utilised. Electromagnets play an important part in modern communications, for example, in telephones and loudspeakers. Huge magnets can be used to drive generators and produce electricity for the national grid. Children will be able to see how the facts demonstrated in their experiments are used in real life.

Part 1 Weather and survival

Wind

Sometimes, the wind blows strongly. Sometimes, we can hardly feel it. How can we measure the strength of the wind?

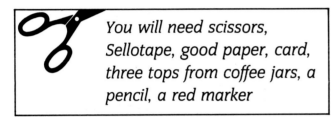

You will need scissors, Sellotape, good paper, card, three tops from coffee jars, a pencil, a red marker

Picture 1 How to make an anemometer

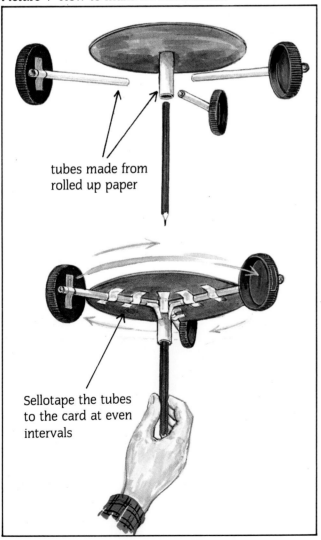

tubes made from rolled up paper

Sellotape the tubes to the card at even intervals

♥ *Experiment: Measuring the force of the wind*

Build the apparatus that these pictures show you. When you have made it, hold it up on a pencil and see that it spins easily. Then stick a strip of red paper on one lid.

Take your apparatus outside. Hold it up on the pencil, so that it can turn freely in the wind. Count how many turns the red marker makes in one minute.

Test the wind strength like this each day for a week. Choose the same time each day.

♥ *Record*

1 Describe how you tested the force of the wind. Draw the apparatus. Make this chart to show your daily results.

	The number of turns in one minute
day one	
day two	
day three	
day four	
day five	

2 How does the wind make the apparatus turn? Try to explain. What would make it turn quickly?

3 Is it important always to measure exactly one minute when you count the turns? Why?

Measuring wind force

An anemometer
You have made a simple anemometer to measure the force of the wind. Weather stations have ones rather like yours. An anemometer has three cups shaped to catch the wind as it blows. The wind pushes each cup in turn and makes the anemometer spin. A strong wind spins it quickly; a gentle wind turns it slowly. Scientists measure the speed of the spin to find the strength of the wind.

What causes wind
Winds are caused by air pressure. Imagine blowing air into a balloon. You will build up a high pressure of air inside. The pressure outside will be lower. If you stop holding the opening closed, the air inside will blow out. You will have made a wind. Air at high pressure always blows into air at low pressure. It keeps on blowing until the pressures are even.

Why air pressure changes
Sometimes, air gets cold – for example, over mountains and icy areas. When air gets cold, it gets thicker and heavier. It

Picture 2 Strong winds can cause great damage

gets a higher pressure. Sometimes, air gets warm – for example, over warm seas and deserts. When air gets warm, it gets lighter. Its pressure gets lower. These differences make winds begin to blow.

❤ To write
1 What is the wind made of?
2 Will wind blow from the cold mountains to the warm desert, or the other way?
3 Winds stop blowing when air pressure is _____ (high, low, even)

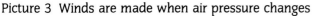

Picture 3 Winds are made when air pressure changes

A weather vane

It is hard to tell the exact direction of the wind. A simple piece of apparatus can help us.

Picture 1 Make a weather vane like this

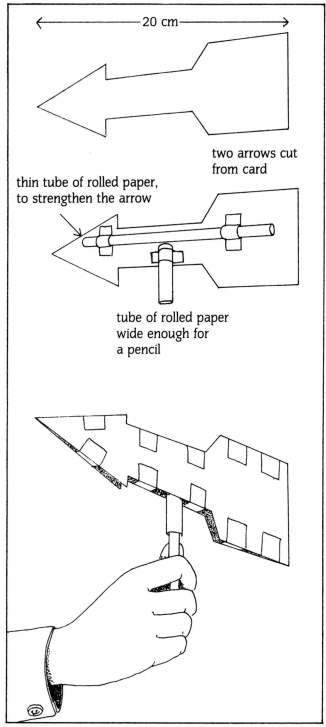

Experiment: Finding the direction of the wind

 You will need scissors, Sellotape, good paper, card, a pencil, a compass

Build the apparatus that the picture shows. Make sure that it spins easily on its pencil. Take it outside, away from buildings and trees. Hold it up in the air and see if it swings into any particular direction.

Hold a compass near the arrow. See if you can work out which direction the wind is blowing from. Record this direction. Do this every day, at the same time.

Record

1 Draw what you have made, and explain what it does. Then make this chart to show your results each day.

	The direction the wind is blowing from
day one	
day two	
day three	
day four	
day five	

2 Does the wind always blow from one direction? Does it tend to blow from one direction more than any other?

3 Who might need to know the exact direction of the wind?

Finding the direction of the wind

Weather vanes

You have made a weather vane. A weather vane shows the direction from which the wind is blowing. It is made so that it swings around easily. One end is larger than the other. It catches more of the wind. The wind pushes this end round, so that the vane turns to point into the wind.

The vane points to where the wind is coming from. We describe a wind by giving the direction from which it blows. So, a northerly wind is a wind blowing from the north towards the south. A south-easterly blows from the south-east towards the north-west.

Picture 2 This tree did not grow straight up because of the prevailing wind

In most places, the wind blows from one direction more than any other. A wind blowing from this direction is called the *prevailing* wind. On high places, you can often see that trees and bushes are all bent over one way. They are leaning away from the prevailing wind.

Picture 3 How air currents move

Weather vanes are put high up, to make them more accurate. Near the ground, wind blows past trees and buildings. Currents of air swirl round them in different directions. A weather vane near the ground is caught in these currents. High up, it is in the main force of the wind.

❤ To write

1 Does it matter which end of a weather vane is big?
2 What is the prevailing wind in your area?
3 An _____ wind blows towards the west. (northerly, easterly, southerly, westerly)

Keeping warm

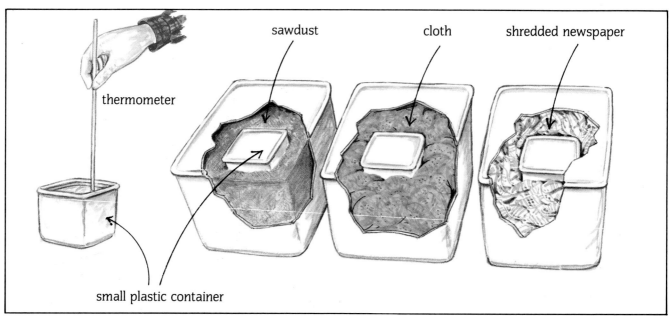

Picture 1 Carry out the experiment like this

Sometimes the weather is bitterly cold. How do animals and people survive?

> You will need warm water, a thermometer, four small containers with lids, three large containers with lids, sawdust, cloth, shredded newspaper

♥ Experiment: How warm things lose heat

Fill a container with warm water. Take its temperature every ten minutes. See how it gradually gets cooler.

♥ Experiment: How to help things keep their heat

Fill three small containers with water as warm as before. Put lids on these containers, and put each one into a larger container. Fill the gap between each small and large container. Fill one with sawdust, one with cloth and one with shredded newspaper. Cover the tops as well as the sides. Then put lids on the large containers.

Every ten minutes, take the temperature of each lot of water. Cover them up again each time. Keep on taking the temperatures for at least half an hour.

♥ Record

1 Describe and draw your experiment. Make four charts like this to show the temperatures.

	10 min	20 min	30 min
Temperature			

2 Do warm things lose their heat? Where does the heat go?
3 Can you help something to stay warm by surrounding it with something else?
4 Do some things keep in warmth better than others?
5 Do animals have anything to keep them warm?

Surviving the cold

Many animals have warm bodies. If they get too cold, they die. Animals need things to keep in their warmth. Things that keep in warmth are called *insulators*. Some of the best insulators in the world are on animals' bodies.

Natural insulators

Pinch your flesh between finger and thumb. You will feel a roll of fat inside. Fat is a good insulator. Many animals have a thick layer of it under their skin. Long-distance swimmers often smear their bodies with grease, which is very like fat. This keeps them warmer in the cold water.

Feathers and fur keep animals warm in two ways. First, they are good insulators. Second, they trap small amounts of warm air. Warm air is a good insulator, and keeps the animal warmer still.

Insulators made by men

Humans do not have fur, feathers or much fat. We keep warm by wearing clothes. Most clothes are made of threads of wool, cotton or other materials woven together. Wool and cotton are good insulators. The threads trap small amounts of air between them.

Picture 2 What helps this seal keep warm?

Picture 3 This Channel swimmer has put grease on to keep herself warm

To write

1 Why do the bodies of animals need insulators?
2 How does the fat (grease) used by long-distance swimmers protect them from cold?
3 Is one thick pullover a better insulator than two or three thin layers of clothing?

Conquering the cold

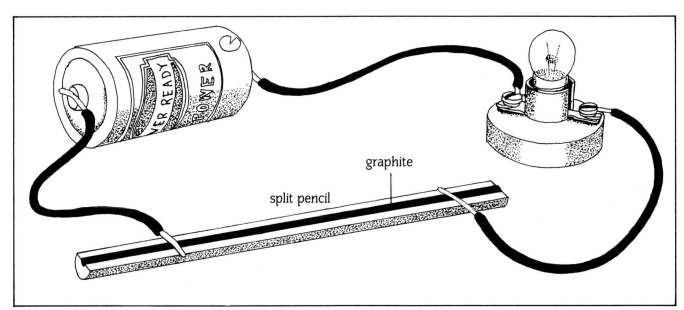

Picture 1 Set up the circuit in this way

People survive cold winters by using electric heaters and lights. How do these work?

You will need a strong battery, a bulb in a bulb holder, three lengths of wire, a pencil split lengthways, a short length of one strand of iron picture wire

♥ Experiment: To resist a flow of electricity

Make a circuit with a battery, bulb and three wires. See how brightly the bulb shines. Then make a gap in the circuit. Put a split pencil by the gap. Hold the ends of the wires close together on the 'lead' of the pencil. (The proper name for this is *graphite*.) You must press hard. See if the bulb lights. Gradually slide the wires apart. Press hard all the time. The bulb will grow dimmer.

Next, hold a strand of iron picture wire across the gap instead. See how this makes the bulb dimmer too.

♥ Record

1 Describe both parts of the experiment. Draw each piece of apparatus. Say what happened in each case.
2 Did electricity flow easily through the graphite? Did the graphite let you control the strength of the electricity? How?
3 Did electricity flow easily through the iron wire? Or did the wire affect it? How?

Electric heaters and lights

The graphite did not let electricity flow quite as easily as the wire. It *resisted* the

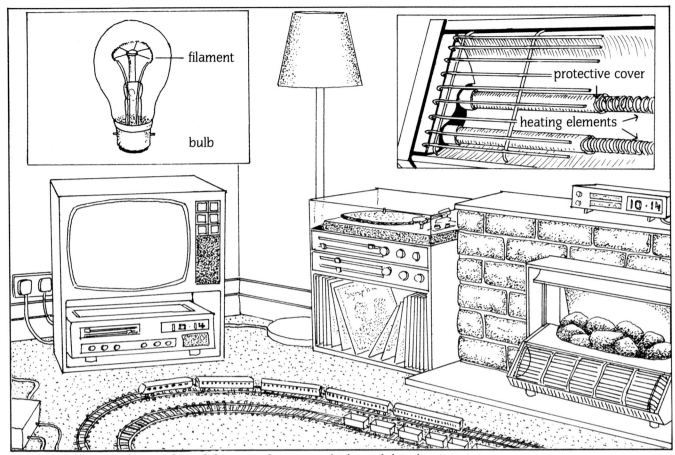

Picture 2 Count up the number of things in the room which work by electricity

electricity a little and made the current weaker. This made the bulb dimmer. When you moved the wires apart, the light grew dimmer still. The current had to pass through more and more of the graphite.

You have learned that electricity flows easily through wire. Some wires resist electricity. When a wire is very narrow, it resists electricity. Imagine a lot of people trying to squeeze through a narrow tunnel. This is like an electric current going through a narrow wire. The wire resists the current.

Fires and lights

When a wire resists electricity, the electricity affects the wire. It makes it hot. Sometimes, it makes it glow. If you had had a powerful enough battery, your iron wire would have grown hot. It might even have glowed.

Electric fires and lights use this effect. The *elements* in fires are made of wire. The wire used has high resistance to electricity. The current makes the wire glow with heat. An electric bulb has a very thin wire inside, called a *filament*. This has strong resistance to electricity and it produces light.

♥ To write

1 What sort of wire will carry electricity without heating up?
2 What else at home is heated by resisting a current of electricity?
3 What could you cut with a hot wire stretched tight?

Rain

Some places seem to get more rain than others. Can we know this for sure? Can we measure how much rain falls?

You will need a plastic bowl, a wide container, a narrow jar, card, scissors, silver foil or polythene, glue, masking tape

Picture 1 How to make a rain gauge

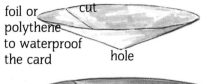
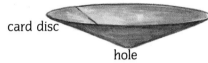
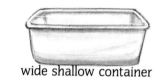

♥ Experiment: Collecting and measuring rainwater

Make a rain collector, as the picture shows you. Take it outside to an open space. Dig a hole to stand it in, to make sure it stays steady. Otherwise, put heavy stones round it.

Each day, at the same time, lift the lid off the bowl. You will see how the container inside has collected rainwater. Pour the rainwater into your narrow jar. Measure and record how many centimetres of water there are.

You have made a simple *rain gauge*.

♥ Record

1 Describe and draw what you have done. Copy this chart so that you can show your results.

		day 1	day 2	day 3	day 4	day 5
How many centimetres of rainwater we measured	10					
	9					
	8					
	7					
	6					
	5					
	4					
	3					
	2					
	1					
	0					

2 Do you think that some places get more rain than others? If so, why? Do some kinds of places get more rain?
3 Why do we bother to measure rainfall?
4 Why did you pour the water into a narrow jar to measure it? Why must you always use the same-sized jar?

Why we measure rainfall

Some places get a lot of rain. Some get a little. It is useful to know which is which. For example, records of rainfall help farmers decide what to grow. If a place has lots of heavy rain, wheat will not grow. But rice might grow well. Weather stations all over the world have rain gauges and keep records of rainfall.

How rain is made

In fine weather, water evaporates into the warm air. It turns into water vapour. The warm air can carry the vapour for thousands of miles. Sooner or later, the air gets cold. Then the water vapour turns back into ordinary water. It falls as rain.

Mountains near the sea usually get lots of rain. Air has to rise high as it blows over mountains. Far above the warm ground, it gets colder. Its water vapour turns into rain. If this air has blown in from the sea, it will have collected lots of water from the sea. It will make a lot of rain.

♥ To write

1 Where is Britain's heaviest rainfall? (Use an atlas to work out the answer.)
2 Why must a rain gauge stand in an open space?
3 Water vapour forms when water _____ . (cools, evaporates, condenses)

Picture 2 The water cycle

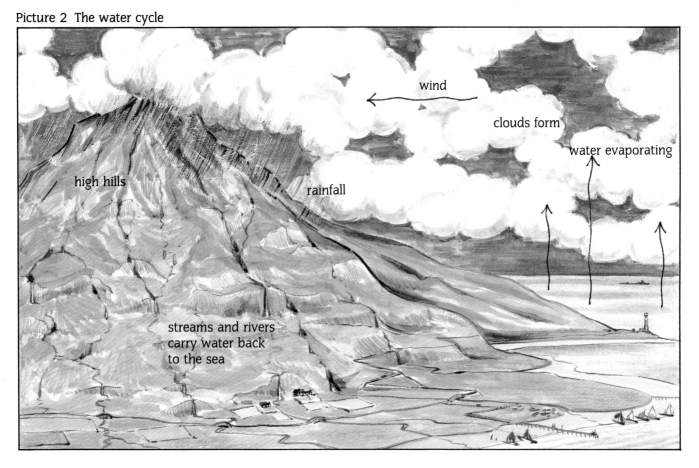

Safe in bad weather

People have often been lost in bad weather. Many of them have found their way back to safety by using a compass.

You will need two strong bar magnets with matching ends marked, thread, paper, scissors, colours, a compass

♥ Experiment: To show that magnets turn to point north

Cut four arrows from paper and colour them like your magnet. Dangle the magnet on thread, in four different places in turn. Each time, wait until it settles. Mark its direction by placing one of your arrows under it. Then use a compass to check the direction of your arrows. See if the magnet pointed north each time.

♥ Experiment: To show that magnets affect each other

Put two magnets on the desk, end to end. Touch the blue end of one to the red end of the other. The magnets will pull together. Next, touch the two red ends together. The magnets will move apart. (Your magnets may be marked in another way to show which ends match.)

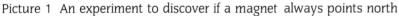
Picture 1 An experiment to discover if a magnet always points north

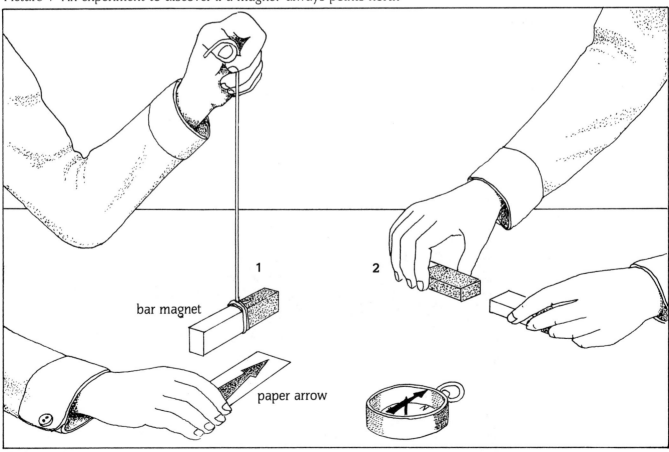

Record

1 Describe what happened when you dangled the magnet. Did one end always point in the same direction? What did this experiment show you about compass needles?

2 Describe what happens when you put two magnets together. Explain the two things that can happen.

A magnetic compass

Some parts of magnets are stronger than others. These are called the *poles*. They look the same but are not alike. One pole always wants to point north. The other wants to point south. A compass works because its needle is a magnet with a pole that always points north.

When two north-pointing poles of magnets touch, they push apart. They *repel* each other. When a north-pointing pole gets near a south-pointing pole, they pull together strongly. They *attract* each other. 'Like' poles repel each other; 'unlike' poles attract each other.

Why magnets point towards north

Picture 3 The Earth is like a huge magnet

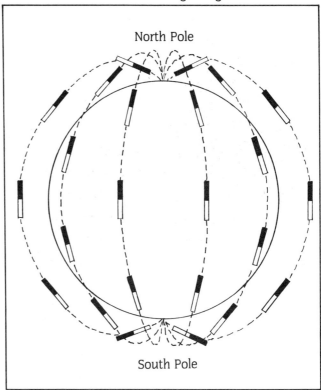

Picture 2 Testing the poles of a magnet

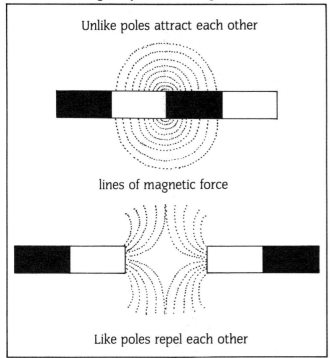

Magnets point north because the Earth itself is like a huge magnet. It has two magnetic poles. One is in the north; the other in the south. These poles are very powerful. The Earth's magnetic north pole pulls at the 'unlike' poles of smaller magnets. It tries to swing them to point north. The magnetic needle in a compass is fixed so that it can swing easily.

To write

1 How would you find the matching ends of two magnets if they were not marked?

2 Magnets repel each other when their poles are _____ (alike, unlike)

17

Weather and survival

Picture 1 Find five things in this picture which help people survive in bad weather. (Look at their clothes, buildings and equipment.)

People have learned how to survive in bad weather. We build insulated houses that stay cool in summer and warm in winter. We use air conditioning in summer and electric heaters in winter. When we travel in bad weather, we wear special clothes to insulate our bodies. We use compasses on land and sea to find our way when bad weather 'blinds' us. We know what weather each season will bring, so we can prepare for it.

Sometimes, bad weather can come suddenly. Winds can uproot trees and smash buildings. These winds, travelling faster than 120 kilometres an hour are called hurricanes. At sea, they make huge waves that can swamp ships.

In the past, people could not be sure when bad weather was on the way. When the storms began, it was often too late to find safety, especially at sea.

An invention called the barometer changed this.

A barometer

A barometer measures changes in air pressure. Changes in air pressure cause winds, so a barometer can tell us when a wind is likely.

The most usual kind is an *aneroid barometer*. An aneroid barometer has a container with sides that can move in and out. There is no air inside the container, and a spring holds the sides apart. The sides are connected to a pointer, which moves when the sides move. Strong, or high air pressure pushes the sides in. This moves the pointer one way. Weak, or low pressure lets the spring push the sides out. This moves the pointer the other way. The pointer shows how high or low the air pressure is.

High pressure usually means that settled air and good weather are likely. Low pressure means that strong winds and bad weather are likely.

Barometers and compasses are important equipment in ships.

Picture 2 An aneroid barometer

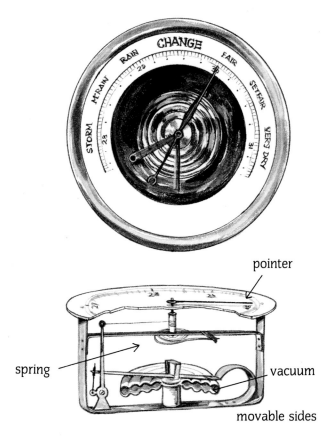

Weather Stations

Men have learned to make careful records of the weather – rainfall, temperature, the amount of water vapour in the air, wind and air pressure. These records are made at weather stations all over the world. Scientists collect reports from the stations and put them together. They build up a pattern of what the weather is doing. They see how the weather pattern is changing, and what is likely to happen.

Picture 3 A weather map of Britain. It shows areas of high pressure and low pressure

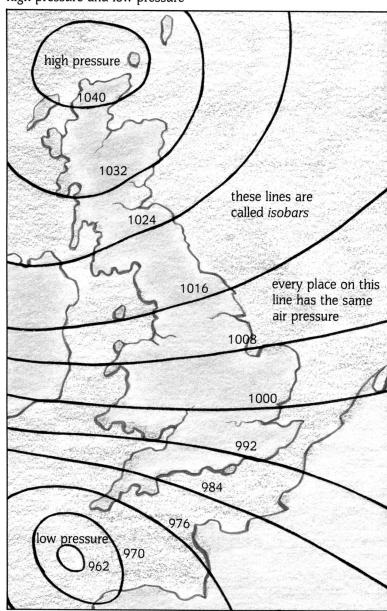

Part 2 Light and how we use it

How we see

Eyes and cameras are alike in some ways. Experimenting with a camera can teach us about our eyes.

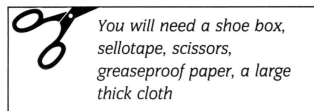
You will need a shoe box, sellotape, scissors, greaseproof paper, a large thick cloth

 Experiment: How a camera works

Take a shoe box and seal it so that it lets in no light. Then cut one end from the box. Glue greaseproof paper over this end. At the other end, make a small hole. You have made a pinhole camera. Take your camera to the window. Use a large cloth to keep daylight from your eyes. Hold the cloth over your head and the box. Point the hole in the camera at the scene outside. Look through the end with the greaseproof paper. See if you can see a picture on it. If not, make the hole a little bigger. Something about the picture may surprise you.

 Record
1 Draw and describe your pinhole camera.
2 Say what kind of a picture the camera made. What was surprising about it?
3 What went into the camera through the hole, to make the picture?
4 In your camera, the picture was made on greaseproof paper. What is the picture made on in a proper camera?
5 How are your eyes like a camera? How are they different?

The picture in a camera

Your pinhole camera worked because light went into it. Suppose that you pointed your camera at a tree. Light from the tree went to your camera. Some of it went through the hole to the paper at the back. Light travels in straight lines. So light from the top of the tree arrived at the bottom of the paper. Light from the bottom of the tree arrived at the top of the paper. This made the picture of the tree appear upside-down.

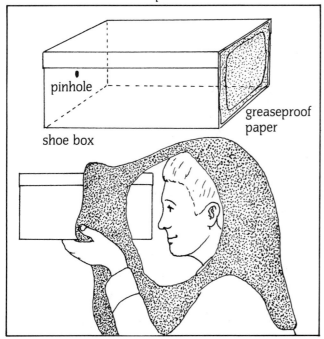
Picture 1 How to make a pinhole camera

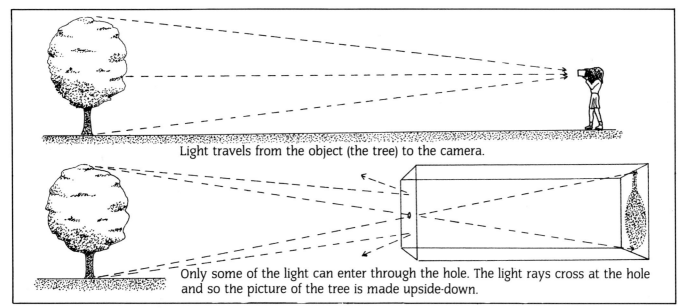

Picture 2 Using a pinhole camera

The human eye

An eye is shaped like a ball. It has an opening at the front, called the *pupil*. This lets light into the eye. The light travels to the back of the eye. It falls on the *retina*. When light touches the retina, the retina sends a signal to the brain. The retina gets an upside-down view, like your camera. But the brain sorts out the signals for us. It makes us see things the right way up.

♥ To write

1 Does it matter if the picture in a proper camera is upside-down?
2 Write a sentence describing what the retina is, and what it does.
3 The _____ stops us from seeing things upside-down. (retina, brain, pupil)

Picture 3 The human eye works rather like a camera. Why do we see things the right way up?

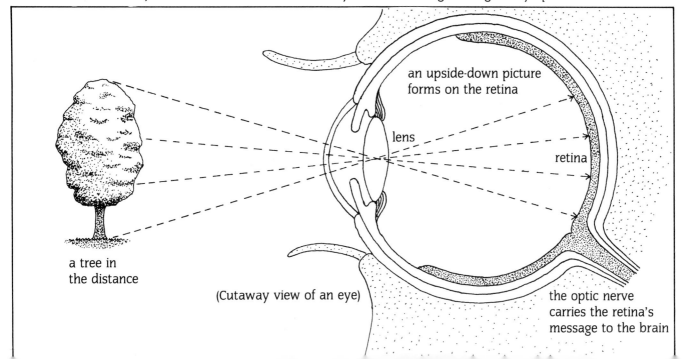

Reflections

What are reflections? How do mirrors make them?

 You will need a mirror, card, a marker pen, a toy person

 Experiment: Reflections

Look into a mirror and see your reflection. Touch your right eye. See which eye your reflection seems to touch.

Print your name on a card. Put the card in front of the mirror and see the reflection. Try to write your name to make the reflection the right way round.

Stand a toy person in front of a mirror. See how close the reflection seems. Now move the toy away from the mirror. See how the reflection seems to move away too. See if the reflection always seems to match the toy's distance from the mirror.

 Record

1 Imagine that someone has never seen a mirror. Describe for him or her what a mirror is.

2 Describe your experiments and draw them. Explain what you found out about mirrors.

3 Make a list of things that can act as a mirror. How are all these things alike?

4 Could a mirror make a reflection if there were no light?

5 Do mirrors really make new copies of things? Or do reflected things only seem to be there?

Light

We see things because of light. Light shines on the things around us – on trees, houses, books and so on. Some of it bounces off these things and comes to our eyes. It shows us that the trees, houses, books and so on are there.

Not all things reflect the same amount of light. Some things reflect most of the light they get. This makes them seem shiny. Polished metal makes a good mirror. Modern mirrors are made of glass with a special silver paint on the back.

Picture 1 Experimenting with reflections

Picture 2 Why does this building reflect the view?

Reflection

Here is how the mirror made a reflection of your toy person. Some light touched the toy and some of it bounced off. Some of it bounced straight to your eyes. This showed you the toy. Some of the light bounced from the toy to the mirror. Because mirrors reflect light so well, the mirror bounced this light to your eyes, too. So you saw a second picture of the toy. This second picture came from the direction of the mirror. So your eyes thought they saw a toy in the mirror, too.

To write

1 Make a list of things which reflect light. Say whether they reflect well, fairly well or not well. Is their colour important?

2 Mirrors _____ light well. (refract, change, reflect, stop)

3 When do things give no reflection at all? How can you tell that there is no reflection?

Picture 3 How a reflection is made

Seeing round corners

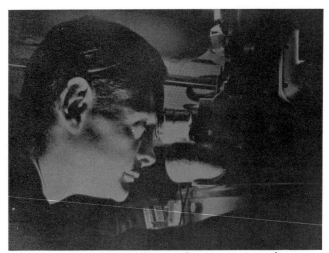

Picture 1 Periscopes help people to see around corners and over things

A periscope lets a submarine's captain see above the waves. It would let you see over a wall that is taller than you.

You will need card, scissors, a set square, ruler, pencil, Sellotape, two small mirrors, a torch

Experiment: Reflecting light from a mirror

Find a dark place, and shine a torch at a mirror. See how the mirror reflects the light. Try turning the mirror a little. The reflected light will change direction. Move the mirror to different angles. See how the reflected light changes direction each time.

Experiment: Using mirrors to see past obstacles

Study Picture 2 carefully. Build the apparatus that it shows you. Make sure that the mirrors are at exactly the correct angles. You have made a simple periscope.

Picture 2 How to make a periscope

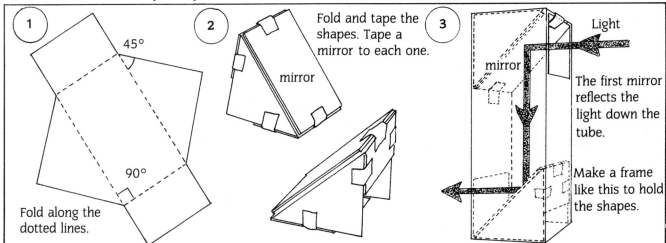

Picture 3 The rear-view mirror in a car allows the driver to see what is happening behind

Hold the periscope upright behind your desk. Look into the bottom mirror. See how the periscope lets you see over the desk.

❤ Record

1 Describe how a periscope is made. Draw the inside of your periscope.
2 Try to explain how it works. Does it work by reflecting light? Does it matter how the mirrors are arranged? Why?

Reflecting light

Mirrors reflect light. When light meets a mirror head on, it is reflected straight back. But when it meets at an angle, it bounces off at an angle. You found this with your torch and mirror. Each time you changed the angle of the torchlight, the reflected light changed direction to match.

Mirrors let us see round corners by reflecting light at an angle. Imagine that a tree is hidden round a corner. Light from the tree cannot reach our eyes. But we can use a mirror to reflect light from the tree. If the mirror is at the correct angle, it will reflect this light to our eyes. Then we can see the tree.

A periscope uses this idea. It has a mirror at the top and another at the bottom. They are set at careful angles. Light goes into the periscope at the top. The top mirror reflects the light down to the bottom one. The bottom mirror reflects the light out of the periscope. It goes to the eyes of the person using the periscope.

❤ To write

1 What do you notice about the angles of the mirrors in a periscope?
2 When you shine light at an angle to a mirror, does it (a) bounce back to the torch? (b) bounce away at the same angle? (c) bounce off in any direction?
3 What would you find a periscope useful for?

Strange reflections

If you look into a shiny spoon, you see a strange reflection of your face. The other side gives another strange reflection. At fairgrounds there are often mirrors which give odd reflections. We say that the reflections are *distorted*. How do they get distorted?

Picture 1 A distorted reflection!

 You will need two flexible mirrors

♥ Experiment: How to make distorted reflections

Hold two mirrors together, flat, in front of you. Look at your reflection. Now turn the mirrors slowly outwards. Watch your reflection. Next, turn the mirrors slowly inwards. Watch your reflection.

Hold one flexible mirror in front of you. Start to bend the mirror. Bend the edges towards you, so that the middle bulges away. See what happens to your reflection.

Next, bend the edges away from you. The middle will bend towards you. See what happens to your reflection.

♥ Record

1 Explain how you used mirrors to make distorted reflections. Draw pictures to show what you did. Try to draw how your reflection changed.

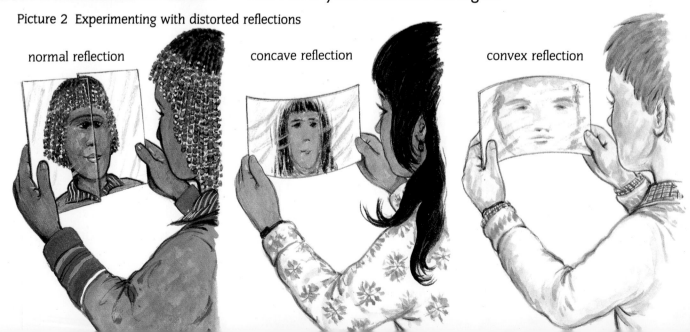

Picture 2 Experimenting with distorted reflections

2 What kind of mirror gives a distorted reflection?
3 Does it make any difference which way a mirror is bent? Does it make any difference how far it is bent?
4 Explain why spoons give distorted reflections.

Flat and curved mirrors

When light from your face meets a mirror, it bounces back. A flat mirror normally reflects light straight back to us. But if mirrors are at an angle, the light bounces off at an angle. When you put your two flat mirrors at angles, they both reflected light in different directions. This made a confusing picture for your eyes.

Now think of your curved mirrors. Each part of a curved mirror is curved slightly differently from the rest. (It helps to imagine a lot of flat mirrors put together in a curve.) Each part of the mirror has a different slant. Each part of the mirror reflects light at a different angle. So a curved mirror does not make a reflection like that of a flat mirror. It gives a distorted reflection.

Concave and convex mirrors.

Some mirrors are curved in a bowl shape. A mirror curved like this is called a *concave* mirror. It reflects light inwards, to a point. (The point is called the *focus*.)

A mirror curved the opposite way is called a *convex* mirror. It spreads light outwards.

A spoon has a concave side and a convex side.

To write

1 Why is it a good idea for a car's rear view mirror to be convex?
2 Mirrors for shaving and putting on makeup are sometimes slightly concave. Why?
3 A _____ mirror spreads light outwards. (broken, flat, concave, convex)

Picture 3 How convex and concave mirrors reflect light rays

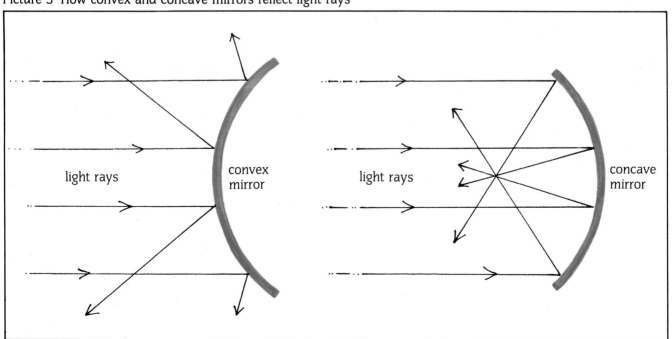

Magnifying

Picture 1 A magnifying glass makes things look larger

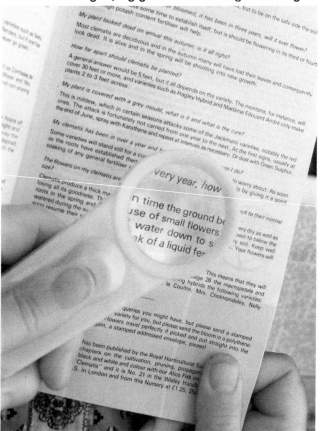

How does a magnifying glass make things look bigger?

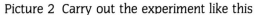
You will need a magnifying glass, a torch, a piece of paper, a book

♥ Experiment: What does a magnifying glass do?

Put a book in front of you. Look at the print through the magnifying glass. See how big you can make it seem. Move the glass up and down between your eyes and the book. Find if it works best at a certain distance. Look at things far away, too. See if they seem larger. **Never look at the sun through a magnifier. In a moment you could be blinded for ever.**

Find a dark place. Shine a torch through the magnifying glass. Hold a sheet of

Picture 2 Carry out the experiment like this

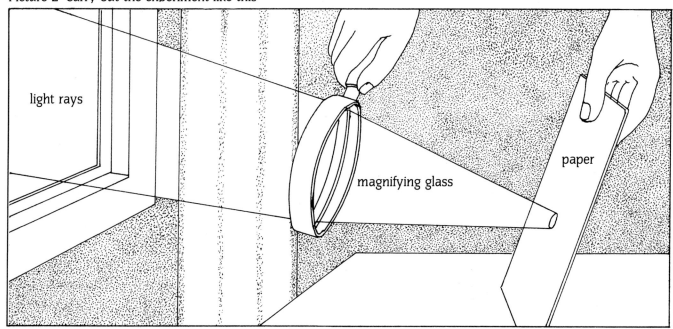

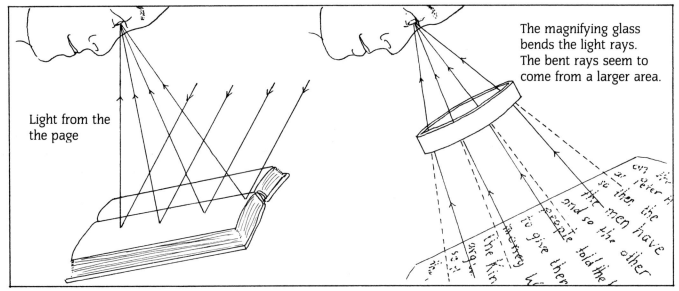

Picture 3 This diagram shows you how a magnifying glass works

paper on the other side. See the circle of light on the paper. Make it as small as you can by moving the magnifying glass. See if it works best at a certain distance from the paper.

❤ Record

1 Explain what a magnifying glass does.
2 Describe your magnifying glass. Feel the glass in it. See if it feels curved. Try to draw what shape it would have if you cut it in half.
3 Describe how you shone light through the magnifying glass. Say if you managed to change the size of the circle of light.

Lenses

A magnifying glass is a kind of *lens*. A lens is usually made of glass or plastic. Glass and plastic affect light which goes through them. They change its direction. We say that they *refract* light. Normally, glass refracts light only very slightly. We do not notice it. But a lens is specially shaped to refract light in a useful way.

How your magnifying glass worked

Light went from the book in all directions. Some rays of light went through the magnifying lens to your eyes. The lens bent this light. (Just as your magnifying glass bent your torchlight into a smaller circle.) This made the light rays seem to come from different places than where they really came from. It made them seem to come from a larger area. This made the letters on the page seem larger than they really were.

Lenses need to be at a certain distance to work best. Otherwise the refracted light does not seem clear. Things look blurred. Then we say that they are 'out of *focus*'.

❤ To write

1 What part of the eye would be badly damaged by the sun's rays?
2 What does an astronomer have in common with a stamp collector and a person with poor eyesight?
3 Name three things which must be in focus if they are to work properly.

Coloured light

Picture 1 Making coloured light

Colour televisions use coloured light to make their pictures. You can discover how they do this.

 You will need a torch, white paper, cellophane or tissue paper (red, green and blue)

Experiment: Making coloured light

Take a torch to a dark place. Set white paper in front of it. Hold red cellophane in front of the torch. (Red tissue paper will do.) See what colour falls on the paper. Change the cellophane for different colours. Try green. Try blue. You have made coloured light.

Experiment: Mixing coloured light

You have made a few colours. See if you can make more. Put red and blue cellophane together in front of the torch. See what colour falls on the screen. Try different pairs of colours together. Keep note of what you do, and what colours you get.

Picture 2 Mixing coloured light

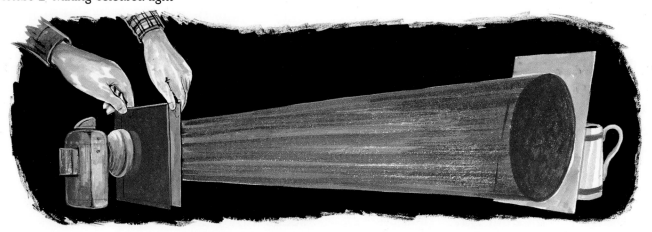

♥ Record

1 Describe how you made coloured light. Then describe how you mixed colours to make new colours.

2 Copy this chart into your book. Use it to show the results of your experiment.

	red	blue	green
red			
blue			
green			

The colour chart will help you to spot the colours you made.

3 Can light be of many different colours? What happens when different colours are mixed?

Ordinary light and coloured light

Ordinary light is a mixture of light of all colours. Your coloured cellophane let you see the colour in ordinary light. You used red, green and blue cellophane. Red, green and blue are called the *primary colours* of light. You can mix these to make any other colour.

Mixing colours

Colour televisions use this idea of mixing primary colours. The inside of the screen is covered with thousands of tiny dots. Some are red. Some are blue. Some are green. These dots light up when they get the right signal. One electric signal lights the red dots. Another signal lights the green dots. Another signal lights the blue dots. To make a coloured picture, different signals are sent to different parts of the screen.

The television can make other colours. To make yellow, the signal is given to light the red and green dots only. The blue dots are not lit. Your eyes see the red and green light mixed together. This mixture appears yellow.

♥ To write

1 Give some examples of how we use coloured light.

2 The primary colours of light are _____ , _____ and _____ . (yellow, red, brown, blue, violet, orange, green, purple)

3 How does a television make colours on the screen?

Picture 3 Coloured lights are used for fun in discotheques

Light and how we use it

Pictures from light

Imagine leaving a newspaper in the sun. Sunlight will turn it yellow. If you leave a shape on the newspaper, it will make a white patch where no sunlight reached the paper. There will be a simple 'picture' of the shape.

Photographs

Some things change much faster than newspaper when light touches them. We call them 'light-sensitive'. Camera films are coated with a light-sensitive mixture. When we click open the shutter of a camera, light goes in and touches the film. It leaves a picture on the light-sensitive film.

We cannot see this picture straight away. We have to wash the film in special stuff called developer. This affects the parts of the film that were touched by light. It makes the film change colour. Slowly, the picture appears.

Negatives and positives

Most cameras make 'negative' pictures. This means that the colours are the opposite of what they should be. You can see this in the picture. We have to shine light through these negatives on to more light-sensitive paper. This makes a new picture, called the 'positive'. This is the right way round.

Some cameras, called Polaroids, make a positive picture straight away.

Lenses

Cameras take clear pictures because they have a lens at the front. We can often move the lens backwards and forwards. This lets us focus light exactly on the film.

Picture 1 A negative picture (left) and a positive one (right)

Picture 2 Telescopes make things in the distance look larger

Picture 3 An astronomical telescope for studying the sky

Refracting telescopes

If we use two lenses, we can magnify things very much. Simple telescopes have a lens at each end. They let us see things that are far away. Sometimes, both lenses are shaped like your magnifying lens. This kind, that curves outwards in the middle, is called a convex lens. Sometimes, one of the lenses curves in at the middle instead. This kind is called a concave lens. But whatever kinds are used, they refract light to make things seem very large and clear.

Reflecting telescopes

A reflecting telescope has a concave mirror at the end. This 'catches' light from far away. It reflects it to a mirror inside the telescope. The mirror reflects this light, through a lens, to the eye of the person using the telescope. Sometimes, a camera is fixed to the eyepiece instead, to take photographs through the telescope.

Scientists use huge reflecting telescopes to study the stars.

Part 3 Magnetism and electricity

Magnetism from electricity

We can turn some magnets on and off. How is this?

> You will need a magnetic compass, two metres of wire, masking tape, a strong battery, a large iron nail, paper clips

♥ Experiment: To show that electricity affects magnetism

Take a compass. Wind a wire round it several times, as the picture shows. The wire should be in line with the north-south compass needle. Fix the compass and wire to the desk with tape. Connect the wire to a battery. Watch the needle. See if anything happens. Swap the wires round on the battery, so that the current moves in the opposite direction. Watch the needle again to see what it does.

♥ Experiment: How to use electricity to make a magnet

Take a large iron nail. Wind wire round it in a coil. The coil must be tight and neat. Connect the coil to a strong battery. Place your iron and coil by some paper clips. See what happens.

Take one of the wires off the battery. See what happens to the magnetism. Then put the wire back, and see what happens.

Picture 1 You can use electricity to make a magnet

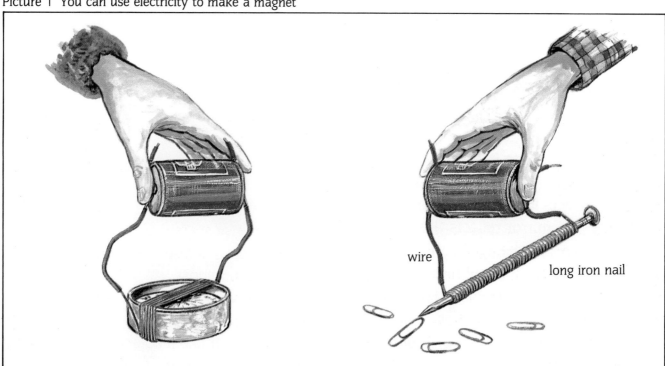

34

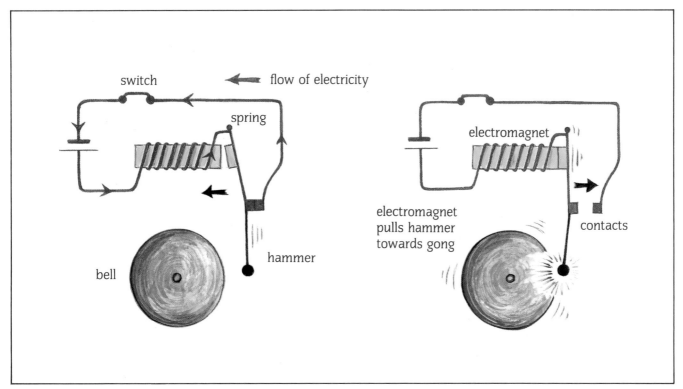

Picture 2 This diagram shows how electromagnetism is used to work a door bell

♥ Record

1 Describe and draw the experiments.
2 Did you show that electricity affected your magnetic compass needle? Is there a connection between electricity and magnetism?
3 Did you manage to make a magnet by using electricity? How was this magnet different from an ordinary bar magnet?

Electromagnets

Electricity produces a magnetic effect when it travels along wire. It produces a *magnetic field*. This magnetic field is usually very faint.

It can be made stronger by bending the wire into a *coil*. Such a coil has a special name. It is called a *solenoid*. You made a solenoid. A solenoid can behave like a normal bar magnet, with a north pole and a south pole. Which pole is at which end depends on which way the electricity is going through. When the electricity is stopped, the magnetism stops too.

The magnetism of a solenoid is made much more powerful if there is iron inside it. The iron becomes magnetised by the coil. Its magnetism is added to the magnetism of the solenoid. Such an arrangement is called an *electromagnet*. Electromagnets are useful because their magnetism can be turned on and off by turning their electricity supply on and off.

♥ To write

1 When electricity flows through wire it makes _____ (noise, movement, magnetism, bends)
2 Say how you would make a very strong electromagnet.
3 Could we make a solenoid by winding a coil around a pencil?

Sound from electricity

You are going to learn how we can use electricity to make sound.

 *You will need some wood, a large iron nail, cardboard, some paper clips, a drawing pin, a strong battery, a metre of wire.*

 *Experiment: How to make sounds from electricity*

Make the apparatus shown in the picture. The paper clips above the nail must be almost touching it.

Connect the coil to the battery. See how the paper clips are pulled down to the nail, and make a click. Disconnect the coil from the battery so that the clips spring back to where they were. See if you can make a series of clicks by connecting and disconnecting the battery.

Record

1 Describe the equipment you made, and draw it. Explain what happened when you used it.
2 Explain why the paper clips clicked down onto the nail.
3 Why did the paper clips have to be metal?

Sound

Your coil and nail made an electromagnet. You turned it on and off by connecting and disconnecting the battery. You were probably able to make a series of clicks as the electromagnet kept pulling the paper clips to the nail and then letting them go. People have used sounders like yours to make messages in morse code. This is explained on page forty.

The electromagnet made the paper clips hit the nail with a click. This sound

Picture 1 Arrange your experiment in this way

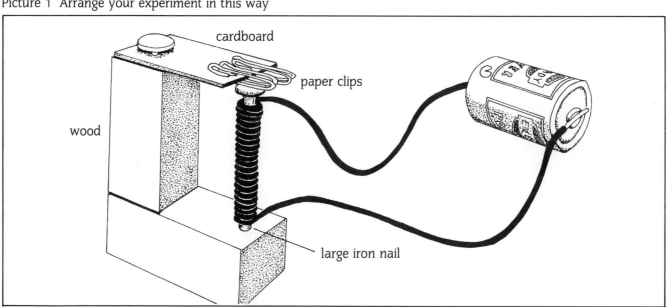

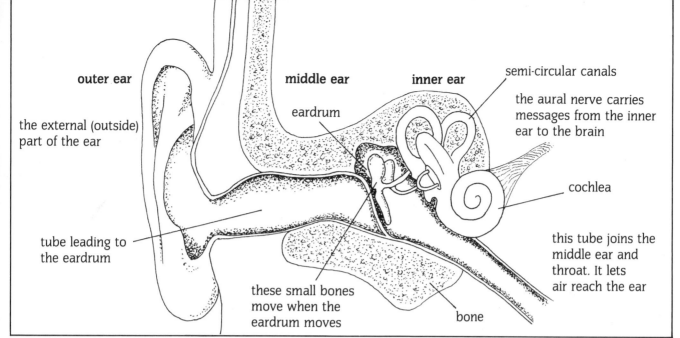

Picture 2 The human ear (a cross-section)

travelled to your ears through the air. When the paper clips hit the nail, it made 'ripples', or 'waves', in the air. These *sound waves* travelled to your ears. A special piece of tight skin in your ears felt these waves in the air. This skin is called the eardrum. Your eardrum made your ear send a message to your brain, and that is how you heard the click.

Sound waves

The idea of waves in the air may sound strange, yet it is not. Think of a pond. If you throw a stone into the middle, you will see ripples, waves, spreading out from where the stone hit the water. The same kind of thing happens to air whenever something hits something else. You cannot see waves in the air like you can in water. But your ears can feel them, and send messages about them to your brain. This is what sound is – waves travelling through the air.

 To write

1 Name three things which make sound from electricity.
2 Describe how covering your ears stops you hearing well.
3 Do you think that sound waves are all of the same size? What would happen if they were?

Picture 3 This shows how sound waves travel through the air to your ear

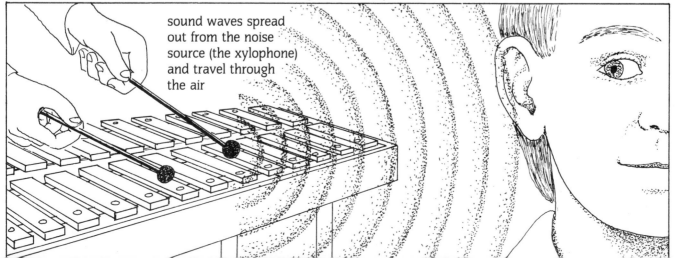

Electricity and safety

Picture 1 Electricity is very powerful and can be dangerous. Do not play with it

We use electricity in our homes that is strong enough to kill us. The electricity in power lines is thousands of times more powerful still. How do we make it all safe?

> *You will need a battery, a bulb in a bulb holder, three wires, a short wire, small pieces of paper, card, chalk, string, plastic, rubber, a matchstick*

♥ Experiment: What will electricity travel through?

Take a battery, bulb and three wires. Join them together to make a circuit. Make sure that the bulb lights brightly. Then make a gap between two of the wires.

See if you can bridge this gap to make the bulb light again. First of all, place a wire across the gap. See if the bulb lights.

Next, put other things across the gap instead of wire. See if the bulb will light. Here are some ideas: rubber, a matchstick, paper, card, chalk, string, plastic. Use only a small piece of each.

Picture 2 An experiment to find out what electricity will flow through

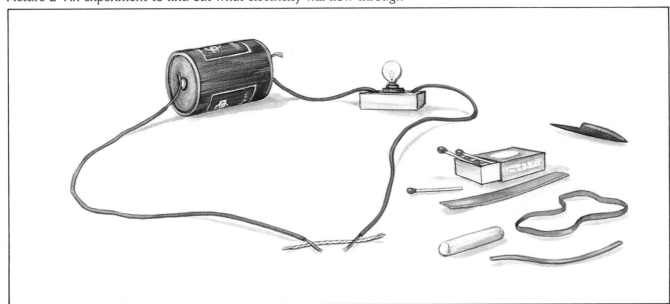

♥ Record

1 Draw the circuit that you made. Describe how you tested different things in it. Say what you found with each one.
2 Does electricity flow better through wire than anything else?
3 Are there any things which will not carry electricity at all?

Conductors and insulators

The bulb lit only when you used wire. Electricity goes through wire. It can travel through other things, but wire is best. Just as a good swimmer moves easily through water, electricity flows easily through wire. We say that wire conducts electricity well. Most metals are good *conductors* of electricity.

Electricity cannot move so easily through other things. Imagine swimming in milk. You would move forward, but the milk would *resist* you. Some things allow electricity to flow, but resist it. They weaken the strength of the electricity.

Non-conductors

Electricity cannot go through some things at all. It is like trying to swim through concrete. These things are called non-conductors, or *insulators*.

Rubber is a good insulator. So are some plastics. Electricity will not normally travel through rubber or plastic. Plugs and sockets are made of rubber or plastic, and electric wires are wrapped in rubber or plastic. When people repair things carrying electricity, they wear special rubber gloves

Picture 3 When people work with electricity they use special tools and clothes

and use tools with plastic handles. These things keep them safe from electric shocks.

Never play around with electricity. Electricity from wall sockets is powerful enough to kill you. You should only use batteries for your experiments.

♥ To write

1 A conductor lets electricity _____ . (vanish, swim, increase, flow, decrease)
2 Light switches used to be made of metal. Why was this dangerous?
3 You are not allowed to have plugs, sockets and ordinary switches in a bathroom. Why not?

Switches

We can use *switches* to send messages with electricity.

You will need a battery, a bulb in a bulb holder, three wires, a switch, a morse (or telegraph) key

Picture 1 You can use electricity to send a message

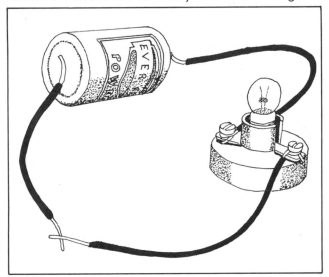

♥ Experiment: Controlling electricity by a switch

Connect a battery, bulb and three wires as the picture shows. Touch the two loose ends of wire together, and see how the bulb lights up. Move the wires apart, and the bulb will go out.

♥ Experiment: Sending a message in code

Try flashing the light on and off quickly. Make long flashes and short ones. Then study the code. Each letter of the alphabet has some dots and dashes by it. Flash the letter 'A' in code. You do this by making a short flash of light and then a long one.

See if you can spell your name in code, and then try to send a simple message.

If you have a switch or a morse key, connect it into the circuit. See how it makes it easier to control the electricity.

Picture 2 The Morse Code alphabet. People used Morse Code to send messages before the telephone was common. Is Morse Code still used today?

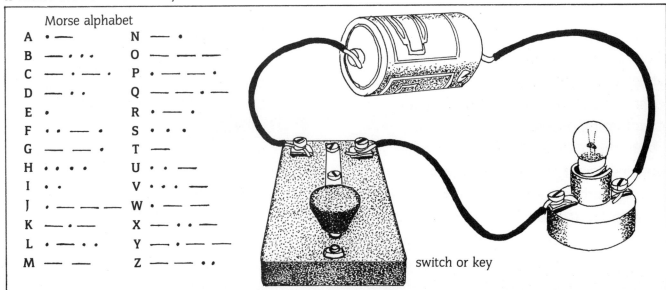

Record

1 Describe how you made a simple switch in your circuit. Then explain how you used the switch to send a message.
2 Say why we use switches with electricity.
3 Explain what morse code is, as if someone had never heard of it.

Controlling electricity

Electricity flows if it can travel back to where it started. It needs a complete *circuit* of wire. If there is a gap, the electricity stops. You made a gap by moving wires. This was a simple switch, which let you turn the electricity on and off.

Electricity is often powerful enough to kill us. Switches to control strong electricity are carefully made. They are made so that we do not have to touch the wires carrying electricity. But they use the same ideas as your switch. When you use any electrical switch at home, the switch moves a piece of metal inside. This metal opens or closes a gap between the wires inside.

The telegraph

Over a hundred years ago, people used telegraph keys to send messages along wires. The telegraph key was a switch that could be tapped quickly. It sent long and short bursts of electricity along wires, as you did when you spelt out your name. The wires were connected to a buzzer or bulb, miles away.

To write

1 What different kinds of switches have

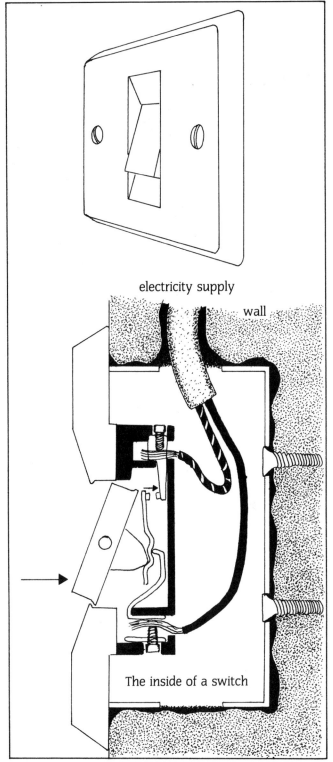

Picture 3 This diagram shows how a switch works

you got at home?
2 Why do we need switches for powerful electricity?
3 Could a telegraph message be sent without wires?

Electricity from a magnet

Electricity can be used to make magnetism. Can we use magnetism to make electricity?

You will need long lengths of wire, a magnetic compass, a strong bar magnet, paper, scissors, masking tape

♥ Experiment: Using magnetism to make electricity

Take some wire and coil it around a compass, in line with the compass needle. This will be a current indicator, like you used for page thirty-four. If electricity starts to flow through the wire, the compass needle will jerk. Tape the current indicator firmly to the desk.

Make a paper tube, large enough to let a bar magnet slide into it. Wind a long length of wire around the tube.

Now join the coil to the current indicator. Keep the coil well away from the compass. Next, take a bar magnet and move it in and out of the coil. As it moves, the compass needle will jerk. You are making electricity flow through the wire.

Notice one thing. If you hold the magnet still in the coil, it will not cause any electricity. It is only when you move the magnet that it causes electricity.

Picture 1 Using magnetism to make electricity

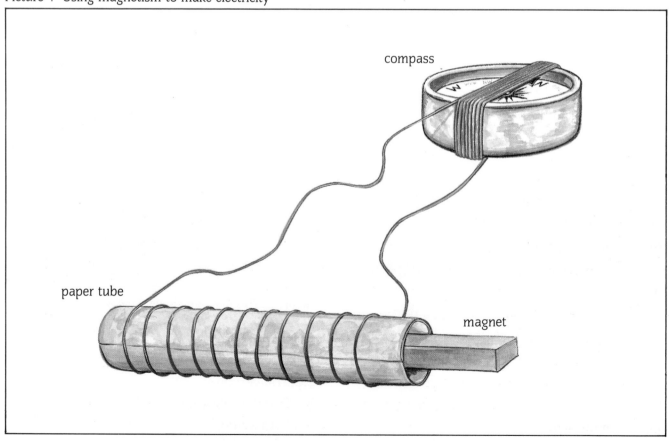

Picture 2 These huge generators are used to make electricity in an hydro-electric power station

♥ Record

1 Describe your experiment and say what happened. Draw the apparatus.
2 We make lots of electricity to work our machines. Does this experiment give you a clue about how power stations make electricity? How do you think that they do it?

Generators

This way of making electricity by moving a magnet in wire was discovered in 1831 by Michael Faraday. Since then, people have learned to make machines using this idea.

These machines, called generators, produce large amounts of electricity. Large, strong magnets are put inside huge coils of wire. These magnets are made to spin round inside the wire. Sometimes, the magnets are made to move by having strong currents of water pushing them. Sometimes, they are connected to powerful jets of steam. As the huge magnets spin, they produce powerful electricity in the wire. This electricity comes to our homes along other wires.

♥ To write

1 A proper current indicator shows when electricity is flowing. What else will it show?
2 What is the smallest generator that you can think of? (Clue: two wheels)
3 To make electricity, generators have to _____ . (stay still, be large, spin, have electric current)

Picture 3 Current indicators (ammeters) – they measure the strength of the electric current

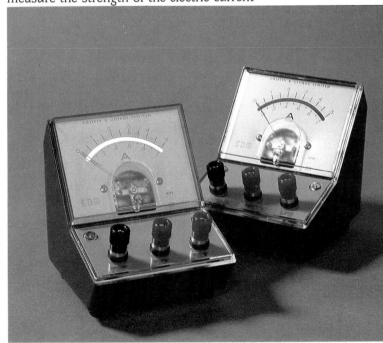

Weak and strong electricity

Does it matter how we connect batteries and bulbs?

 You will need a battery, three bulbs in bulb holders, plenty of wires

Experiment: Two ways to arrange a circuit

Join the battery and bulbs together as the pictures (1) and (2) show. Make sure that the wires are connected firmly. See how brightly the lights shine each time.

Record

1 Describe the two ways of connecting bulbs to a battery, and draw them. Say if either of them made the bulbs shine more brightly.
2 What makes the bulbs shine more or less brightly?
3 Did the way the wires were arranged affect the strength of electricity in each bulb?
4 If you had added more bulbs to the top arrangement, would they have shone even more dimly?

Picture 1 Arrange your bulbs in series like this

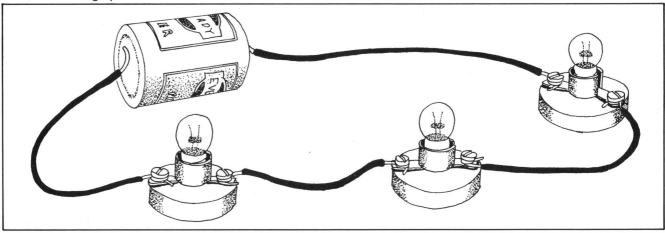

Picture 2 Arrange your bulbs in parallel like this

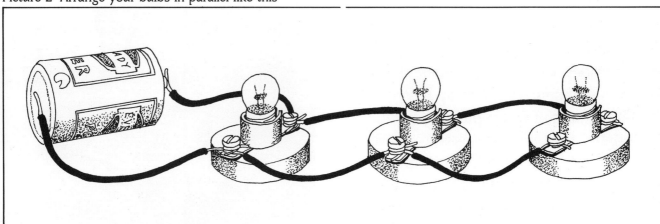

Series and parallel

In your first arrangement, you put the bulbs in line, one after another. You put them *in series*. Electricity had to travel through the first bulb before it could get to the others. If you had taken out the first bulb, the others would have gone out. When things are arranged in series, the strength of the electricity is divided among them. The more bulbs that there are, the weaker the electricity that each bulb gets.

The second arrangement is called arranging *in parallel*. Each bulb is joined directly to a main wire coming from the battery. If you take out one bulb, the others keep on shining. Each bulb gets the full strength of electricity. Each bulb shines brightly.

Comparing series and parallel

Each type of circuit has good points and bad points. Bulbs arranged in series do not shine as brightly as they would in parallel. But they do not use up too much electricity. The battery can last a long time.

Bulbs arranged in parallel do shine brightly. But they take a lot of electricity from the battery. They wear out the battery more quickly.

Arranging batteries

You can arrange batteries in series and in parallel. Three batteries in parallel produce the same strength of electricity as one battery. But they go on producing it for three times as long. Three batteries in series produce three times the strength of electricity. But the batteries wear out sooner.

To write

1 What would be the two main disadvantages of having the lights at home in series?
2 Several lights in a long, dark corridor are controlled by a switch at each end. Should the lights be in series or parallel?
3 Batteries last longer when they are _____ . (in series, used a lot, in parallel)

Picture 3 This diagram of a car battery shows you how it works

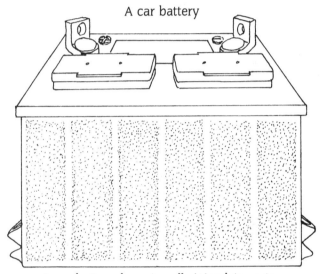

A car battery

A car battery has six *cells* joined in series

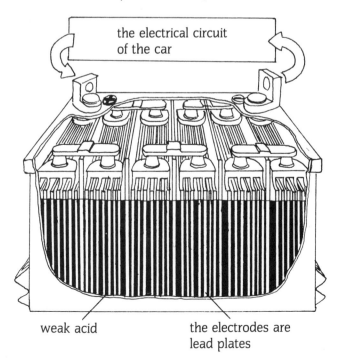

the electrical circuit of the car

weak acid   the electrodes are lead plates

Magnetism and electricity

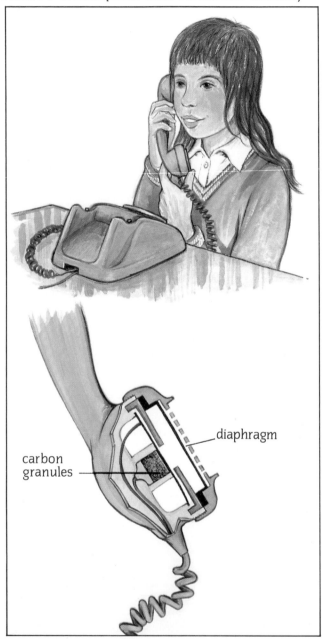

Picture 1 A microphone turns sounds into electricity

The telegraph

You have learned how a telegraph system used electricity to send messages. In the early nineteenth century, many people tried to make a telegraph that worked well. One of the most successful was Samuel Morse, who also invented the Morse Code. The first public long-distance telegraph message was sent in America in 1844. Telegraph systems were powered by batteries like yours. Sometimes, the wires were connected to a special printer. This printed out the Morse Code messages on tape.

Later, a man called Marconi invented a way of sending a message without using wires. His 'wireless telegraphy' sent messages in Morse Code. Later still, we learned how to send the sound of the human voice. To do this, the sound of a voice had to be turned into electricity.

Microphones

Sound was turned into electricity by using microphones. A microphone can turn even complicated sounds like music into electricity.

In a modern microphone there are many grains of stuff called carbon behind a disc. Wire carrying electricity is connected to this carbon. When you speak into a microphone, you make sound waves. These make the disc tremble. As the disc trembles, it pushes the grains of carbon in and out slightly. This movement affects the strength of the electricity in the wire. Loud noises move the disc most, and cause the biggest change in the electricity. Soft sounds hardly move the disc at all, and cause only a weak change. As you talk into a microphone, you send a changing electric current along its wire.

Earphones

Earphones turn this changing electric current back into sound waves. An earphone has an electromagnet in it. This electromagnet has a metal disc in front of it. As the different strengths of electricity go into the electromagnet, they make its magnetism weaker or stronger. These changes in magnetic strength make the metal disc tremble. This trembling makes sound waves in the air. These sound waves are an exact copy of the sound waves that went into the microphone. We hear an exact copy of the sound.

A telephone has a microphone and an earphone inside it, so that you can send and hear a message.

Loudspeakers

We often make the sound waves extra clear by joining the metal disc to a cone of cardboard. The whole cone trembles when the disc does. This moves more air, and makes clearer and louder sound waves. Radios, record players and cassette players all have loudspeakers inside them.

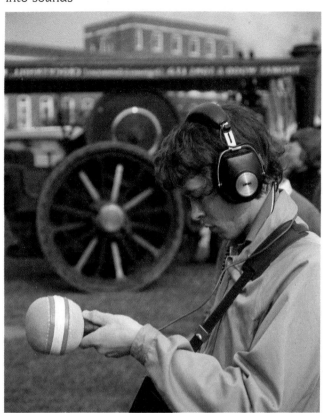

Picture 2 Earphones change the electrical current back into sounds

Picture 3 This diagram shows how loudspeakers make sound clearer and louder

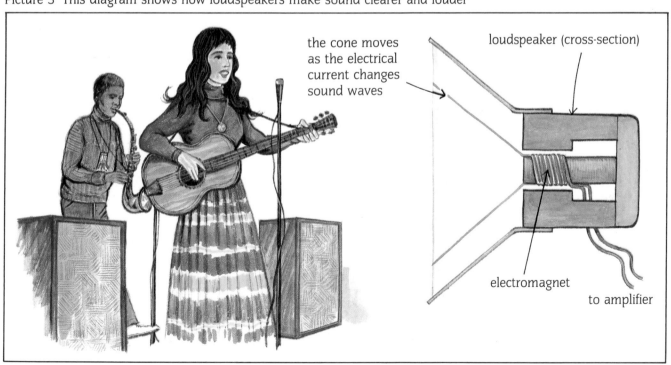

New words

anemometer an instrument that measures the force of the wind
aneroid barometer an instrument that measures the pressure of the air
attract to pull from a distance

circuit an arrangement for carrying electricity, usually made of wires
coil wire twisted into rings that lie next to each other
concave curving inwards at the middle, like a dish
conductor something that carries electricity
convex curving outwards at the middle, like a dish upside-down

distorted changed, altered

electromagnet something that produces magnetism from electricity
element the part of an electric fire that glows with heat

filament the wire coil in a bulb that glows with light
focus the exact placing of a lens to refract light as clearly as possible

generator a machine which makes electricity from magnets moving inside wire coils
graphite the black stuff in the middle of pencils

hurricane a powerful storm wind travelling faster than 120 kilometres per hour

insulator something that stops heat or electricity from getting through

lens something curved to refract light in a special way, usually of glass or plastic

magnetic field the area of magnetism around a magnet or a wire carrying electricity

non-conductor something which will not let electricity flow

poles the most magnetic parts of magnets and of the Earth
prevailing wind a wind that usually blows from one direction
primary colours colours which can be mixed to make all other colours
pupil the part at the front of the eye which lets in light

rain gauge an instrument for measuring rainfall
refract to bend light
repel to push away
resist to make it difficult or impossible for electricity to flow
retina the back of the eye, which is sensitive to light

solenoid coil of wire which acts as a magnet when electricity flows through it
sound waves ripples made in the air when things move or knock against other things
switch something which lets us make a gap in a circuit without touching the wires